2nd Topical Conference on Heat Exchangers 2022

Topical Conference at the 2022 AIChE Spring Meeting and
18th Global Congress on Process Safety

San Antonio, Texas, USA
10 – 14 April 2022

ISBN: 978-1-7138-6821-7

Printed from e-media with permission by:

Curran Associates, Inc.
57 Morehouse Lane
Red Hook, NY 12571

Some format issues inherent in the e-media version may also appear in this print version.

Printed with permission by Curran Associates, Inc. (2023)

For permission requests, please contact AIChE
at the address below.

AIChE
120 Wall Street, FL 23
New York, NY 10005-4020

Phone: (800) 242-4363
Fax: (203) 775-5177

www.aiche.org

Additional copies of this publication are available from:

Curran Associates, Inc.
57 Morehouse Lane
Red Hook, NY 12571 USA
Phone: 845-758-0400
Fax: 845-758-2633
Email: curran@proceedings.com
Web: www.proceedings.com

TABLE OF CONTENTS

640017 Heat Exchangers Future Market : Energy Transition - Net Zero By 2050

Abstract:

Heat Exchanger market is getting ready for energy transition in the decades ahead. The challenge is satisfying the growing population and energy demand with reducing the risks of global warming, using new technologies to improve energy efficiency to have reliable, affordable, and sustainable energy resources and carriers around the world.

Statistic data shows renewable energy supply and demand will continue to increase but it will not be an easy process since it will require more land for production. Storage systems and batteries developments are required, and new infrastructures must be installed. Using fossil fuels in energy market will continue to provide reliable back up resources, therefore the future of energy market will be a combination of renewable energy and clean fossil fuels to reach carbon dioxide emissions to Net Zero by 2050 according to the roadmap specified by International Energy Agency (IEA) issued in May 2021 and United Nations Climate Change Conference, COP 26. Fossil resources will also stay in the global market for non-energy purpose, chemical and petrochemical industry using new technologies such as Carbon Capture Utilization and Storage.

Heat Transfer equipment designers, manufacturers and end users should be aware of the rapid change in supply and demand of heat exchangers due to energy transition to find out how consumers' preferences will change the technical requirements during design and fabrication of exchangers in the new projects. Following topics will be discussed during presentation:

1) Global Warming

2) NET ZERO EMISSION by 2050

3) Heat Exchangers Market, design and Fabrication challenges for:

- Fossil Fuels: Oil and Natural Gas for Energy and Non-Energy use
- Carbon Capture, Utilization and Sequestration
- Hydrogen and Ammonia
- Biofuel

640019 Particle-sCO_2 Primary Heat Exchanger Technology for Moving Particle Thermal Energy Storage Systems in High Temperature sCO2 Based CSP Plants

Abstract:

Renewable energy systems tend to be intermittent with high capital costs and low thermal-to-electric conversion efficiency. In order minimise costs, they would need to operate near full capacity, which is one reason for their limited integration in the electric grid mix. The penetration of renewable energy generators on the electric grid is significantly improved by integrating long duration energy storage systems necessary to match production with demand. Various energy storage technologies are being considered and developed. These include thermal systems using sensible storage materials (molten salt, solids and water) and phase change materials (organic, inorganic, liquid air energy storage); mechanical storage systems (compressed air energy storage, pumped energy storage, gravity and flywheels); electrochemical storage or batteries (Li-ion, Flow, Lead Acid) and chemical energy storage (hydrogen, ammonia, methanol).

In this paper, we discuss a heat exchanger technology that allows integrating solid particle based thermal storage systems with an advanced concentrated solar energy systems. Concentrating solar power (CSP) with thermal energy storage has the potential to provide dispatchable renewable energy with increased capacity factor. CSPs have evolved from oil based trough systems (~400^{0}C) or molten salt-based (~565^{0}C) Steam Rankin cycles to systems targeting higher than 700^{0}C turbine inlet temperature by coupling with a high-efficiency supercritical carbon dioxide (sCO_2) power cycles.

Solid carbo, ceramic, or sand particles suitable for high operating temperatures have the advantage of being low cost and abundant heat storing medium with superior thermal properties including high solar absorbance. These small particles, sensible thermal energy storage media, can work with various receiver configurations and help meet low levelized cost of electricity and cost of energy storage. However, there is a need to have a reliable and low cost, high effectiveness heat exchanger between the thermal storage media (particle) and working fluid (sCO2). One technology that is being developed and tested for the particle-to-sCO_2 primary heat exchanger (PHE) is diffusion bonded plate sub-assemblies stacked with gaps (enclosed with side bars) creating a moving particle bed. The bonded sub-assemblies use plates with chemically etched fluid flow channel network suitabile for high temperature and pressure sCO2 fluid and popularly used in sCO2 power cycles as recuperators and coolers. The narrow slots between the bonded sub-assemblies provide compact moving particle bed with higher heat transfer contact.

In comparison with the conventional alternatives that uses tubular fluidized bed designs, moving particle monolithic plate designs result in higher overall heat transfer coefficient with added advantage of being easier for system integration and packaging due to their compactness. Demonstration scale test results and future solid side heat transfer enhancement technologies will also be presented in this paper.

SafeHX®

Buffered Heat Exchangers

For Increased Safety And Temperature Control Of Sensible Fluids.

Nelson Clark, Clark Solutions, São Paulo, Brazil
Vitor Sturm, Clark Solutions, São Paulo, Brazil

Introduction

Sulfuric acid is the most produced chemical in the world. In 2021 it is estimated that the world has produced around 260 million Metric tons[i] of this critical chemical for the world food safety.

The production of sulfuric acid, from Sulphur burning processes or other routes, is a exothermic process. While a substantial amount of energy is recovered a large amount of energy is still rejected and wasted.

Finding safe routes for recovering the currently wasted energy and reducing environmental pollution has always been a growing concern of the industry.

During many years Clark Solutions has pursued a solution for those problems. Ultimately this effort led to the development of SafeHR[ii] and SafeHX[iii] technologies, that truly allow the recovery of this wasted energy without risk for plant operators.

The latest stage of this development process culminated in the SafeHX design that, once finally devised, allowed the finding of other uses, benefits and potential applications for this technology.

This paper intends to briefly present the progress of the development of these new technologies as well as present the results achieved in industrial and prototype installations.

Background

Sulfuric acid cooling has always been a major safety and operational continuity challenge to the sulfuric acid industry.

The vast majority of sulfuric acid plants, particularly those that burn sulfur, generally recover energy through a sequence of exothermic processes on the gas side. The very exothermic sulfur burning process can produce steam in quantities comparable to the plant sulfuric acid production.

Moreover, the conversion of SO_2 to SO_3 generates a significant quantity of heat. As the temperature rises throughout the conversion process. Heat must be removed to give room for more conversion as the system approaches equilibrium. This energy can be recovered enhancing plant vapor quality.

Finally, the production of acid by the reaction of sulfur trioxide and (SO_3) and water (H_2O) releases heat, approximately 15 kW per metric ton of acid produced, which traditionally is wasted in the form of cooling water evaporation.

$$SO_{3in\ gas} + H_2O_{(l)in\ 98,5\%wt.H_2SO_4} \rightarrow H_2SO_{4(l)} \qquad \Delta H^o \approx -130\ \frac{MJ}{kmol} H_2SO_4$$

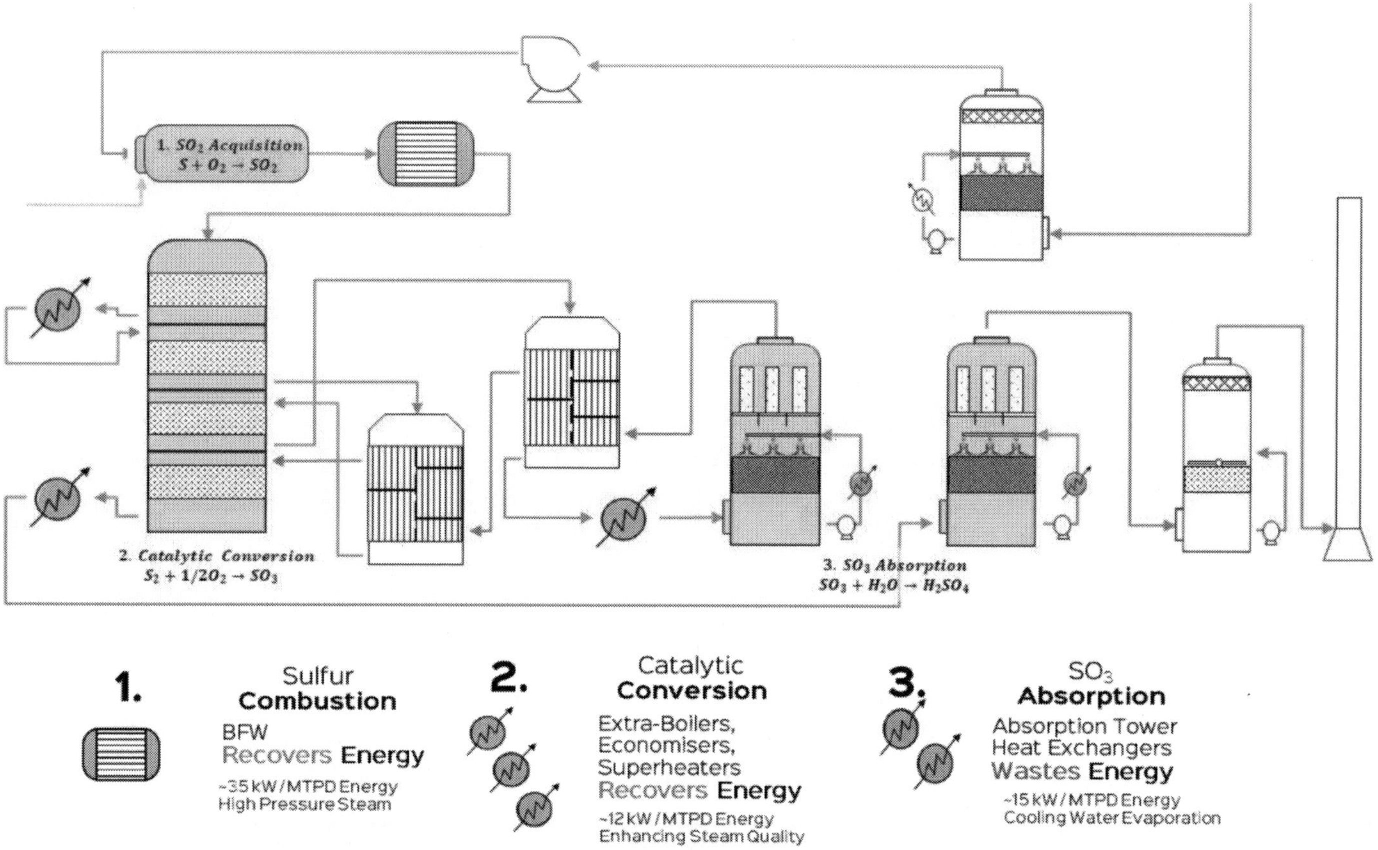

Figure 1: conventional sulfuric acid double contact absorption plant. Emphasis on exothermic processes and energy recovery.

It has been the goal of the industry for almost 40 years, since Monsanto's first patent to a hot acid absorption and heat recovery technology[iv]. Depending on the Sulphur source this process would increase heat recovery by 35% up to 80+%..

The tempting heat recovery process, however, seats on dangerous grounds. While strong acid gets more corrosive as temperature rises, the reaction between it and water releases heat and generates more corrosive hot diluted acid that catalyzes the process leading to a destructive spiral that ultimately will lead to a plant shut down and equipment damages.

$$H_2SO_{4\,(\ell)} \rightarrow H^+_{(aq)} + HSO^-_{3\,(aq)}$$

$$\Delta H^0 = [(\Delta H^0_{f\,HSO_4^-} + \Delta H^0_{f\,H^+}) - \Delta H^0_{f\,H_2SO_4}] \approx -72\ kJ/mol$$

Any high school student knows: "be careful mixing acid and water".

This energy, with few exception, is still being rejected to the environment.

The search for solutions to this problem drove us to engage in a process of replacing water as the acid cooling media. This led to the patenting of 2 (two) technologies:

- SafeHR®: Safe Heat Recovery technology

and

- SafeHX®: Safe Heat Exchanger

In order to present SafeHX, it is necessary to first make a brief introduction to SafeHR.

SafeHR – Safe Heat Recovery Technology

The SafeHR approach to the problem consists of the development and implementation of intermediate fluid trim cooling system.

The fluid, chosen to be inert to acid and water, have good heat transfer properties, as well as other important characteristics like non-flamability, atoxicity, intermediate density between acid and water, among others.

Figure 2: Intermediate fluid unhazardousness.

Once the proper fluid for a given service is selected, the next step is to set up the system. Basically the system consist of a pair of heat exchangers, shell and tube or plate, a pump to drive the intermediate fluid, a coalescer-expansion vessel where any leak fluids will be detected and the piping, valves and instrumentation to operate the system.

The system set, hot acid is cooled by the circulating intermediate fluid which in turn is cooled by heating cooling tower water (if no heat recovery needed), boiler feed water or generating low pressure steam directly.

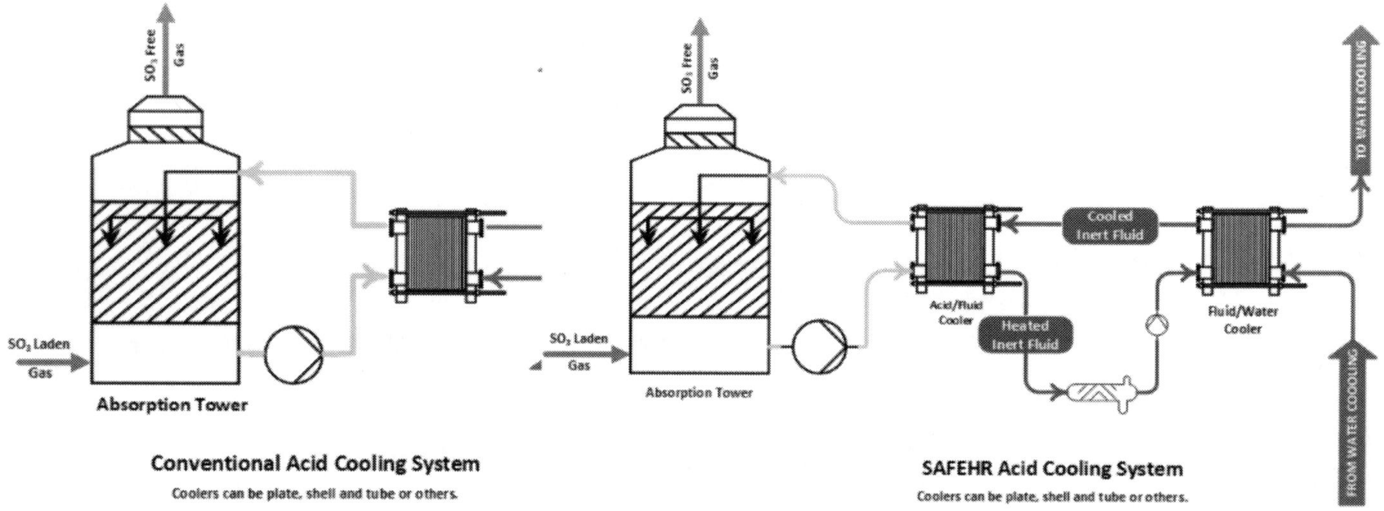

Figure 3: conventional sulfuric acid cooling or heat recovery (left) , SafeHR acid cooling or heat recovery system (right).

The idea behind SafeHR is simple: should a leak occur, the leaking fluid will be collected in the intermediate system avoiding the detrimental acid-water contact and allowing operator to plan the maintenance of the system.

The first industrial installation of this system has been implemented in Brasil, on a 150 metric ton per day sulfuric acid plant. The unit has been started-up in February 2020 and is currently operating.

Figure 4: SafeHR system as installed in an industrial plant. Left, water-fluid evaporator exchanger, center, heat exchangers, liquid liquid coalescer and fluid circulation pump, right absorption tower and acid-fluid heat exchanger.

The process proved successful. Heat recovery and safety achieved design objectives.

Safety has been subject to an actual operational event late 2020. Due to water treatment problems, the fluid/water exchanger failed in October 2020.

Figure 5: water-fluid exchanger failed due to cooling water problems.

At that moment of the incident, sulfuric acid prices were soaring in Brasil and, as planned, the system did not have to be stopped due to the leak. To avoid production losses, estimated in $ 150,000/day in acid production, customer chose to monitor the situation instead fo shutting down.

The plant kept running, with the leaking exchanger, for about 8 (eight) months, when the steam production requirements from the site were reduced and the system has been able to be stopped and maintained (as the acid plant was designed by Clark Solutions it can run with or without the heat recovery system). As a matter of fact, the acid plant is being stopped for maintenance on April, 2022 (statutory requests).

Should the same event happen in a conventional acid cooling or heat recovery system, this a even more delicate case due to the water pressure (7 bar), with hot acid and water would instantly mix and lead to certain plant shut down and potentially dangerous scenario.

SafeHX® – Safe Heat Exchanger

SafeHR proved efficient, reliable and has actually avoided an incident that elsewhere could turn into a serious episode.

Nevertheless, the search for improvement made a new product to bloom: SafeHX.

SafeHR system required two exchangers, intermediated fluid to fulfill both exchangers and the circulation system and all the peripherals to make the system work: pump, expansion vessel, piping valves and instrumentation.

Could this be simplified? How to put all these features in a single exchanger while keeping the safe design?

Everything Inside a Single Shell

The solution to this equation used the properties differences between the fluids: if the boiling point range of the intermediate fluid is within the hot and cold fluids, a piece of equipment can be designed so as the hot fluid boils the intermediate fluid as it gets cooled and the cold fluid condenses the boiling vapors as it gets heated.

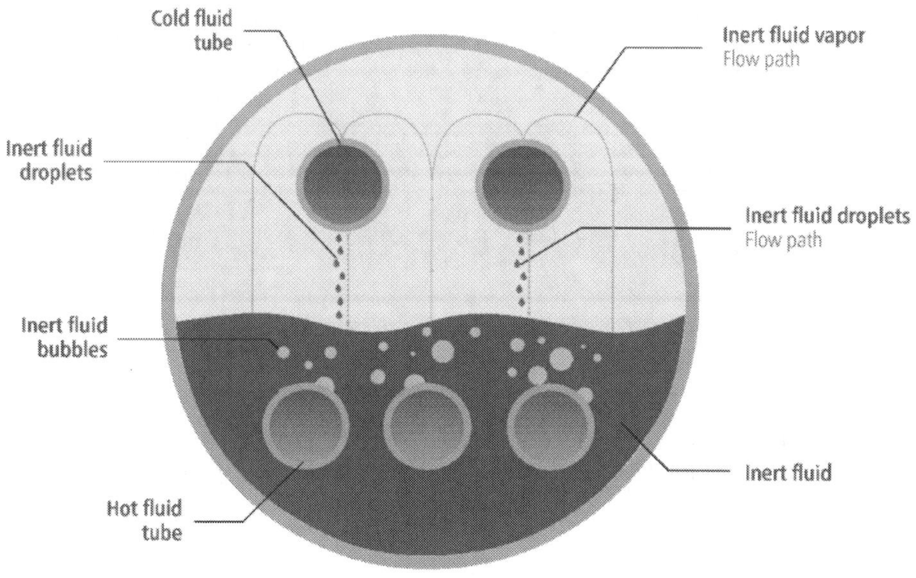

Figure 6: SafeHX operating principle

In the limit, the system has an intermediate fluid contained within the heat exchanger, that is constantly boiling and condensing, indirectly transferring heat from one fluid to the other, but still preventing that they get in contact in the event one side of the system fails.

The simplest design for such system is a shell and tube heat exchanger, with two tube bundles where the process fluids flow inside the tubes and the intermediate fluid boils on the shell side.

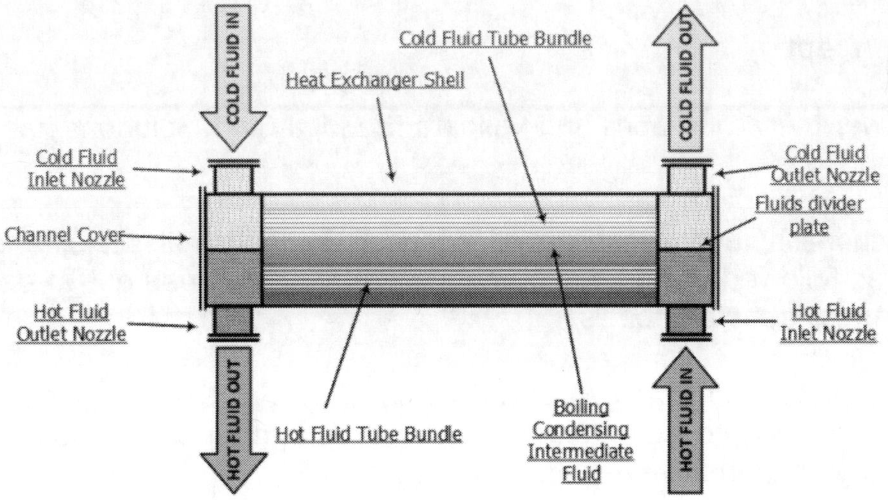

Figure 7: SafeHX heat exchanger arrangment

This design offer the same safety benefits of a SafeHR system, avoinding a potential hazardous situation in the event of a cross contamination of fluids without the equipment, instruments and extra costs required in a more complex situation.

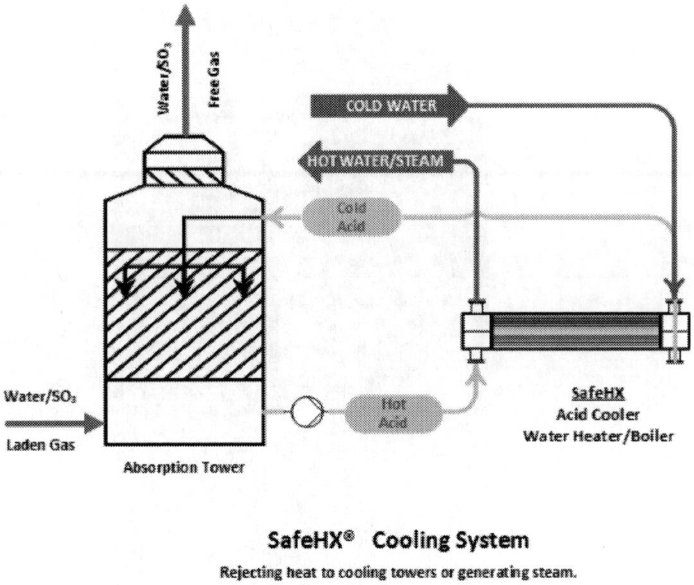

Figure 8: SafeHX flow diagram

The concept looks promising, but a series of questions arise:

- Will the boiling/condensing approach actually work?
- Will the heat exchange between rising vapor and dripping condensed liquid affect the heat transfer process?
- How the coupling of two extra effects, boiling and condensing, will affect the overall heat transfer coefficient?

Testing the concept

In order to answer these questions and validate the idea Clark Solutions has built a pilot unit to try the concept.

To ensure functionality and capability to be used in industrial scale, a clear and direct methodology was followed. A rigorous numerical study was carried out to size the prototype, using heat and mass transfer models.

Using obtained results from this numerical study, Clark Solutions engineering team carried out a mechanical project, an experimental bench setup and an automation system aiming for a validation through quantitative results.

The experiment bench is presented in the Figure 9. In its concept there were: two pumps, two volumetric flow meters, five thermistors and three pressure transmitters. All the bench control and data acquisition were held by the PLC and a supervisor system. The supervisor system was operated through a computer in which bench control was undertaken.

Figure 9: SafeHX prototype.

The basic flow diagram of the bench in Figure 10, shows it has two closed-loop water cycles. One is called the cold-water loop. This loop has a tank, a pump, a flow control valve, one pressure transducer, two temperature transducers and one air radiator, to decrease water temperature. The other is the hot-water loop. This other loop also has a tank, a pump, a flow control valve, one pressure transducer, two temperature transducers but it has a heat resistance instead of the air radiator, to increase water temperature.

The intermediate fluid, a mix of fluorinated hydrocarbons, was selected to be inert to water (and in the case also to acids and bases) and to boil in a 40-80 C range, since cold and hot water would be fed to the exchanger at approximately 30 C and 80 C respectively.

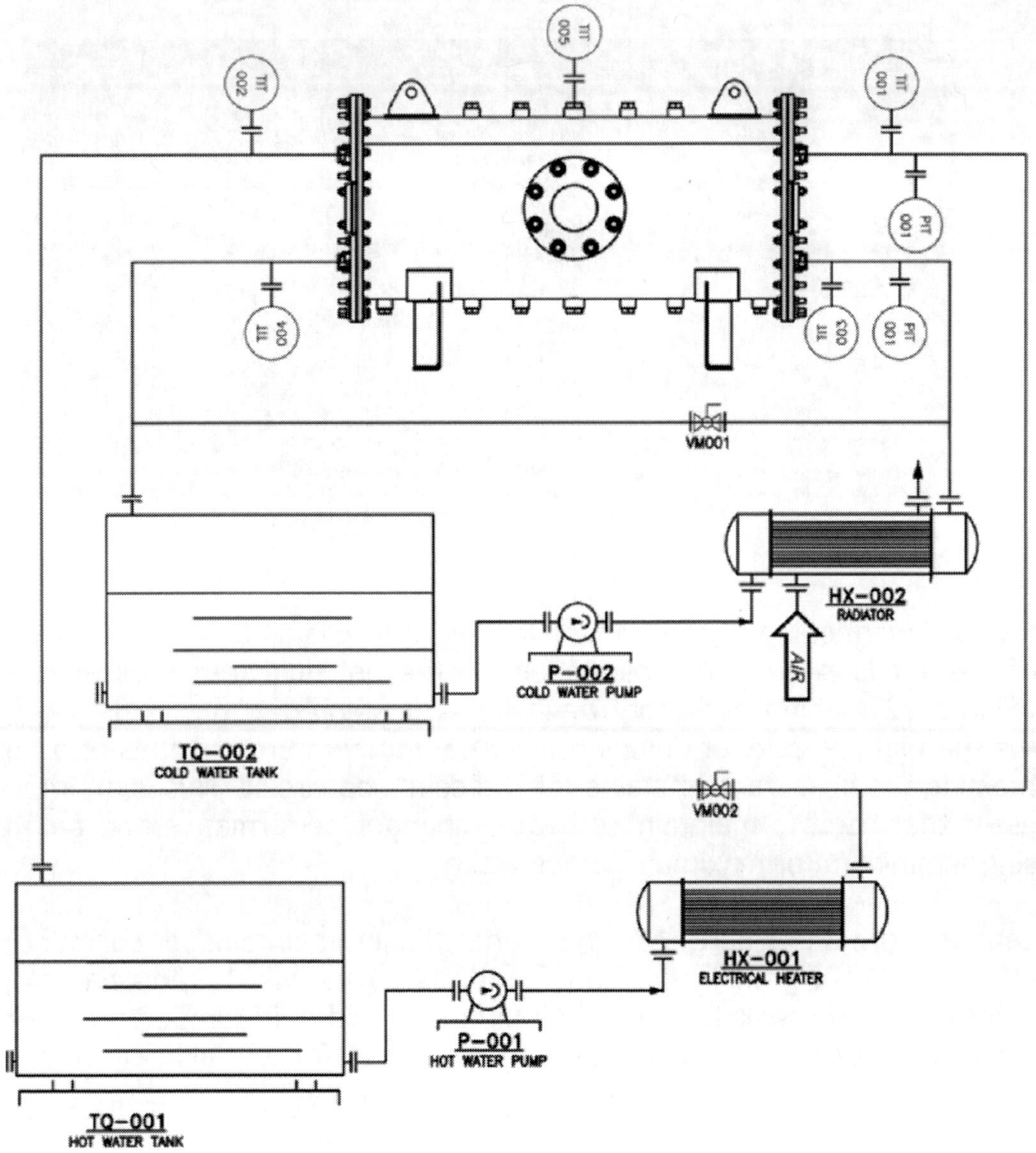

Figure 10: SafeHX prototype basic flow diagram

Experimental Results

After several experimental bench tests to ensure that instrumentation and system were working as expected it was possible to execute some prototype runs to verify its heat transfer capacity.

As a evidence of the heat transfer phenomena, we have the visual of both boiling and condensation processes inside the heat exchanger shell.

The boiling is verified by the froth which hide the submerged tubes, and the condensation is verified by the uniform dripping from the emerged tubes. In figure 11 we see a very disturbed liquid surface caused by vapor bubbles and an intense flow of droplets dripping from the emerged tubes.

Fig. 11: SAFEHX working behavior

It is also worth mentioning the intermediate fluid buffer effect. As the temperature control system of the hot loop generates oscillations in the hot fluid temperature, it is possible to verify in Figure 12 that no oscillatory behavior is presented in the cold loop temperatures. This shows the high capacity of damping eventual temperature changes of a hot loop circuit and maintaining the heat transfer stable for the cold loop circuit. This is an important feature for processes that need to maintain its heat exchanger performance and gain the safety of not crossing a minimum or maximum temperature.

As we can see on the Figure 13, the buffer liquid maintains a constant temperature controlled by SAFEHX® shell pressure setting point, which dominates buffer liquid temperature. Even with oscillations on hot fluid inlet, cold fluid temperatures are stable. The system proves to not only be a safe means for heat transfer, but also an excellent system to control temperature.

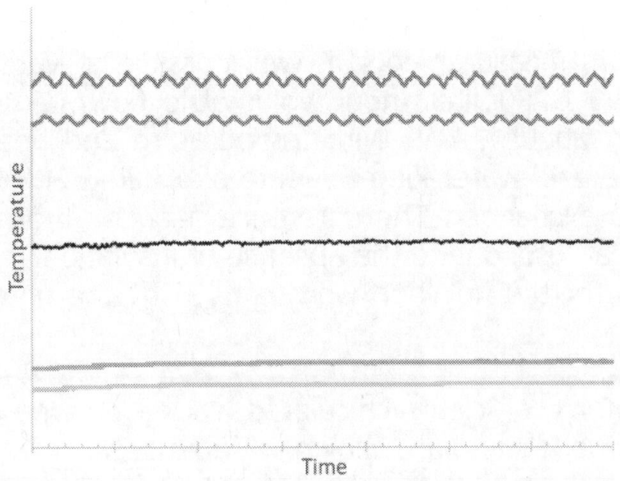

Fig. 12: SAFEHX® experimental data (red line – hot inlet temperature; blue line – hot outlet temperature; black line – buffer liquid temperature; orange line – cold liquid outlet temperature; green line – cold liquid inlet temperature)

An overview of heat transfer can be seen in Figure 10, which further illustrates the dampening effect caused by the buffer fluid-controlled temperature.

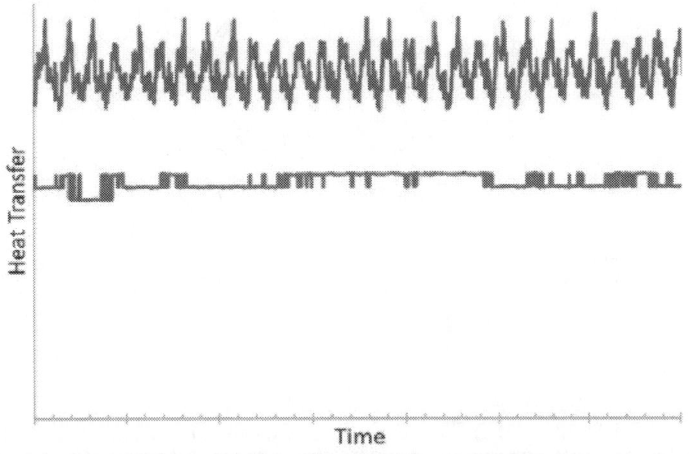

Fig. 13: SAFEHX heat transfer for hot and cold fluid (red line – hot fluid heat transfer to the system; blue line – cold fluid heat transfer from the system)

The system was tested with varying heater output to change hot fluid inlet temperature over time. The excess heat was subsequently transferred to the buffer fluid with some heat lost to the surroundings (not perfectly adiabatic shell as observed from the gap between lines).

This transferred heat from hot liquid to the SAFEHX system can be seen in the red lines of the previous Figure. But the erratic buffer liquid energy intake from the hot loop did not necessarily cause de cold fluid to behave as erratically, this may seem contradictory at first, but this occurs due to internal energy variations of the buffer liquid which dampen system temperature variations. Blue lines have plateaus which are caused by changes in cold liquid flowrate and not differences in cold fluid temperatures.

SAFEHX Performance

After processing sensors measurements, it was possible to verify that the heat exchanger, operating with a 1.000-1.200 liters/hour volumetric flowrate in the hot-water loop, was capable of transferring about 4 kW. With temperature and heat transfer results shown in Figures 10 and 11. The cold-water loop flowrate of 1.000-1.200 liters/hour retrieved roughly 3 kW of heat from the exchanger. The difference between those two heat rates came from the shell convection heat loss, due to the absence of insulation.
The visible oscillations for the hot loop were caused by the on-off control of the hot water loop temperature.

The overall heat transfer coefficients calculated from prototype measured parameters were close to the numerical models and are shown in Table 1.

Further prototype improvements and different sets of operation conditions were already implemented and, just installing shell insulation, shown an increase of approximately 15% on the overall heat transfer coefficient.

As a starting point of the technology, it is seen that we already have a prototype with heat transfer performance for liquids compatible with commercial products and with all the safety and process control improvements already mentioned.

Table 1
SAFEHX Tests Results

Fluid Pair	Conditions	Results
Cold Side Heat Transfer Coefficient	Inert fluid condensation to warm cold water	500 to 1.000 $W/(m^2 \cdot K)$
Hot Side Heat Transfer Coefficient	Inert fluid evaporation to cool hot water	500 to 1.000 $W/(m^2 \cdot K)$
Overall Heat Transfer Coefficient	Continuous evaporation and condensation of buffer fluid	250 to 500 $W/(m^2 \cdot K)$

Applications and Usage

Originally devised to be used as a safe and economical means of cooling and recovering heat from sulfuric acid production, SafeHX can be used in a distinct series of other processes and situations:

- Avoiding Hot and Cold Fluid Mix:

  Sulfuric acid cooling is probably not the only process where a hot-cold fluid mix may create a dangerous or the unacceptable mixture of fluids. Phosphoric acid / water mixtures is another example.

- Fouling fluids:

  A frequent problem is conventional shell and tube heat exchangers is where to place the fouling fluid or the maintenance of shell side fouling. SafeHX does not have a shell side flowing process fluid. Both process fluids flow on the tubes and, thus, cleaning and tube plugging can be easily accomplished by opening the channel covers.

- Buffered heat exchange:

  There are several processes and situations where precisely controlled temperatures are required to avoid fluid decomposition, for instance in the heating of amines. These situations as general rule require the use of multiple heat exchangers to, slowly and steadily provides heat to the amines without exceeding certain tubewall limit.

  SafeHX, by means of properly chose intermediate fluid, will accept high hot fluid temperature and still provide controlled temperature to the cold side. The increase of energy input on the hot side will increase boiling, but not the intermediate fluid vapor temperature, protecting the cold fluid from thermal damage.

Conclusion

SAFEHX® is a new, breakthrough approach, that adds a substantial safety feature to existing or new sulfuric acid plants. The prototype achieved its original objective of safe heat transfer in a compact device as originally intended. But it also benefits heat transfer operations with high fouling rates, operations where cold liquid temperature control is important, and for cases where process liquid loss is costly. All of this achieved with a cost-effective equipment. The prototype is in advanced stage and a comercial version of the product should be available in 2021.

SafeHR® and SafeHX® are trademarks of Clark Solutions.

References

---

[i] Source: https://www.statista.com/statistics/1245226/sulfuric-acid-market-volume-worldwide/, acessed 3/17/2022
[ii] Clark, N.P. (2020) United States Patent No. US10,875,770; São Paulo, BR: U.S. Patent and Trademark Office
[iii] Clark, N.P. (2020) Patent Pending
[iv] McAlister, D. R.; Ziebold, S. A. (1986). United States Patent No. US4,576,813. St. Louis, MO: U.S. Patent and Trademark Office

639415 Preventative, Predictive and Prescriptive Maintenance of Heat Exchanger Networks: What's the Difference in Economic and Environmental Terms?

Abstract:

The process industry is going through a digital transformation and heat exchangers are a key part of it. Traditional preventive maintenance philosophies that seek to clean heat exchangers at regular intervals are more and more replaced with predictive tools that enable the optimisation of cleaning actions. Through predictive maintenance, operators are warned in advance of potential issues (e.g. operating limits being hit) and action can be taken to minimise unexpected shutdowns, loss in production, fuel consumption and environmental impact.

A prescriptive maintenance approach, which still quite new to the industry, goes even further allowing the operator not just to be warned when it is optimal to clean but also which is the optimal method to use and for how long. However, both predictive and prescriptive maintenance approaches hinge on the ability of the software tools to accurately predict fouling in the heat exchangers, its impact on plant operations as well the description of the effects that cleaning actions have on the performance of the units once they are brought back into operations.

In this presentation, we will use a Hybrid Digital Twin and a case study from a real plant to assess the difference between a preventive, predictive and prescriptive maintenance approach to cleaning heat exchangers in terms of overall plant economic and environmental emissions.

637541 One-Dimensional Transient Model of a Heat Exchanger Tube Rupture Using Thermophysical Data: Discharging a High-Pressure Liquid into a Low-Pressure Liquid-Filled Shell

Abstract:

The process industries require heat or cool High-pressure (HP) liquid. The most common method used has been shell and tube heat exchanges. The Low-pressure side (LP) of exchanger which contains a utility fluid is therefore at risk in the event of any leakage from High-pressure (HP) side of heat exchanger. Such exchangers need to be designed concerning maximum pressure generated under the tube rupture event. The heat exchanger designers have to consider the peak pressure value of the shell in the event of tube rupture to select suitable materials for the shell. It has become common practice for the LP side to be designed to withstand a pressure above the operating pressure of the utility fluid. The risk of tube failure could lead to failure of the LP pressure shell and release of large quantities of flammable gas. The LP side, therefore, should be protected against tube failure by fitting relief devices. The consequences of such a failure can cause a catastrophic rupture of the LP side with considerable financial loss and risk to personnel.

API Std 521 (Clause 4.4.14.2.2) recommends that dynamic analysis be performed to determine the overpressure and the required relief flow rate within the protection devices. Dynamic analysis using an appropriate simulation program allows calculation of the transient pressures in the LP side of the heat exchanger following a tube rupture. Dynamic analysis is required to ascertain that the maximum surge pressure that could be reached does not compromise the integrity of the LP side of the exchanger.

The use of one-dimensional models is preferred as their accuracy is fit-for-purpose. The alternative approach is 3D/2D Computational Fluid Dynamics (CFD) which requires significant computing time and higher cost, especially if the model intends to include the upstream and downstream piping of the shell.

Several one-dimensional dynamic models have been developed in the literature (Ennis et al. (2011), K. K. Botros (2015), and A. Harhara and M. F. Hasan, (2020)). However, these models ignored the effects of temperature on their transient pressure calculation. If the released liquid in the shell being in contact with hot shell liquid reaches its superheat limit, the increased heating and vaporization contributes to pressure buildup in the event of tube rupture. Hence, the exclusion of temperature effect seems to underpredict the peak pressure.

This study aims to investigate the suitability of a heat exchanger design for a flashing liquid-liquid system under the case where one tube ruptures and the release of process liquid inside the shell occurs. In the proposed system, the subcooled liquid will flash once entering the low-pressure shell-side. A 1D dynamic model will be developed using Python to perform the transient overpressure analysis which can predict the excessive pressure of the shell in the tube rupture event of the heat exchanger. In this paper, the thermophysical data is a key input and needs to be provided before the 1D simulation. This research explains how thermophysical properties data is generated and employed in the 1D model. In the current work, the effect of temperature and vaporization will also be included.

638972 A Cryogenic Test Rig for Startup Procedures of Main Heat Exchangers in Flexible Air Separation Units

Abstract:

An increasing share of renewable energy sources in the power supply is leading to a higher volatility of the German energy market. One possible way to stabilize the power grid is the flexible operation of energy-intensive processes such as cryogenic air separation. Hence, the Kopernikus project "SynErgie", which is funded by the German Federal Ministry of Education and Research (BMBF), investigates the enhancement of the load flexibility of air separation units (ASUs)[1].

The main heat exchanger was identified as one of the key components for dynamic plant operation because the higher number of startups, shutdowns and load changes can lead to thermal fatigue and decrease the lifetime of this vital plant component. Aluminum plate-fin heat exchangers (PFHEs) are used in these plants because of their high process integration, low production cost and compact design.

Deeper understanding and experimental data from dynamic operation of representative PFHEs are of major interest. Thus, a large cryogenic test rig for dynamic operation of PFHEs was designed, constructed, and put into operation [2,3]. In a cyclic operation scheme, two PFHEs are alternately cooled down to cryogenic temperatures and heated up to simulate startup procedures in ASUs and induce thermal fatigue eventually. Extensive measurement technology, like distributed temperature sensing using optical fibers and detailed strain evaluation, offer interesting insights into the behavior of PFHEs under dynamic operation [4]. The collected data from the test rig is used for validation of a three-dimensional model for thermo-fluid simulation [5] and computational lifetime estimation tools.

Results show severe 3D temperature maldistribution, which results in high thermal stress for this harsh test scenario. Two promising operational measures for lowering thermal stress in dynamic operation of PFHEs were examined at the test rig. Results will provide important information for design and operation of main heat exchangers in flexible ASUs.

Literature:

[1] Klein, H.; Fritsch, P.; Haider, P.; Kender, R.; Rößler, F.; Rehfeldt, S.; Freko, P; Hoffmann, R.; Thomas, I.; Wunderlich, B.: Flexible Operation of Air Separation Units. Chemie Ingenieur Technik 92 (12), 2020, pp. 1921 – 1940

[2] Haider, P.; Freko, P.; Lochner, S.; Reiter, T.; Rehfeldt, S.; Klein, H.: Design of a test rig for the simulation of startup procedures in main heat exchangers of air separation plants. Chemical Engineering Research and Design 147, 2019, pp. 90-97

[3] Fritsch, P., Hoffmann, R., Flüggen, R., Haider, P., Rehfeldt, S. & Klein, H.: A Cryogenic Test Rig for Dynamically Operated Plate-Fin Heat Exchangers. Chemie Ingenieur Technik, 93(8), 2021, 1230–1237.

[4] Fritsch, P., Hoffmann, R., Flüggen, R., Woitalka, A., Haider, P., Rehfeldt, S. & Klein, H.: Distributed Temperature and Strain Measurements at a Cryogenic Plate-Fin Heat Exchanger Test Rig: in production. Chemie Ingenieur Technik, 2021, Pubblicazione anticipata on-line. https://doi.org/10.1002/cite.202100070

[5] Haider, P., Freko, P., Acher, T., Rehfeldt, S. u. Klein, H.: A transient three-dimensional model for thermo-fluid simulation of cryogenic plate-fin heat exchangers. Applied Thermal Engineering 180, 2020, p. 115791

639588 Crude Unit Heat Integrationa Case Study

Abstract:

Crude Unit Heat Integration

A Case Study

Andrew W. Sloley

At a minimum, crude oil will remain the source of transportation energy through 2050, and perhaps beyond. Crude oil distillation is the first step in converting crude into streams for further processing into transportation fuels.

Crude oil distillation is energy intensive – all the desired products are distillates that must be vaporized. Modern units use complex heat integration to recover the energy from condensing the product and reflux and using it to heat the feed. Extensive academic and industry work has gone into optimizing the heat integration for maximum energy efficiency and minimum investment. However, finding the optimum configurations remains complex and time consuming.

This is even more true for revamping units. Revamps add additional constraints to the energy analysis. This may make what appear to be obvious energy integration steps impractical. A case study for a recent revamp is presented that balances multiple factors to find a profitable and more efficient heat integration configuration.

639480 Heat Exchanger Network (HEN) Retrofit for a Sustainable Energy Efficient Design in Aramco's Lube Refinery

Abstract:

Industrial sector is accountable for approximately one-third of world-wide energy usage that relies on fossil fuels as the primary source of energy. A significant percentage of energy consumed is spent to supply utilities for oil & gas facilities so that energy could be engendered for other sectors. The adverse impact of fossil-fuel combustion has instigated efforts to minimize carbon emissions, whilst environmental regulations have a substantial effect on the energy's production cost. Furthermore, it is a competitive world with ever-increasing energy prices, industrial communities are convened to brainstorm over conservation of energy resources and profitability. Energy efficiency should be a foundational strategy to support plans to meet the climate policy goals as well as for being lucrative. Energy efficiency with its integral role in energy intensive refining processes is critically essential to achieve desired energy-saving potential and reduction in CO_2 emissions. Pinch technology imparts a systematic technique for saving energy in processes and on entire sites. This technique helps optimize thermal heat recovery, and immediately became popular as a theoretically elegant yet feasible approach to design Heat Exchanger Networks (HENs). The paper illustrates, through a case study, the 'plant lifetime retrofitability' concept in a standard oil refinery's VDU plant. The study demonstrates the impact of energy demand reduction by identifying retrofit projects for a Vacuum distillation unit (VDU), utilizing retrofit with retrofit in mind approach. The case study's objective is to look for the most inexpensive retrofit projects via arranging each initiative in phases for implementation using the existing topology and available capital investments. Each energy saving' initiative is implemented in a way that does not hinder the future identified initiatives. The analysis was conducted with a commercially available Heat-Int software, with an initial minimum temperature difference (ΔT_{min}) of 34°C with corresponding hot and cold utility requirements of 21.53 Gcal/hr and 17.13 Gcal/hr, respectively. The analytical outcomes disclosed a 2.02 Gcal/hr energy reduction in the vacuum distillation unit (VDU) with simultaneous generation of MP and LP steams of 4.4 and 3.8 tonnes/hr, respectively. The proposed modification can be implemented in three phases, execution of each phase retrofit won't hinder the future retrofits and beyond.

639506 Heavy Crude Oil Processing to Improve Pipe Line Transportation Properties

Abstract:

Rationale

Transporting heavy crude oils through pipeline or by other transportation means is challenging and often requires the petroleum industry to resort to rail transport for inland sources, adding large volumes of diluent, or adding very high energy input, not to mention the extra handling and danger of spillage.

Objective

Describe a novel approach to economically transporting heavy crude oil by pipeline over long distances by reducing the viscosity of heavy crude oils using low-concentration chemical addition and integral mechanical treatment. The treated heavy crude oils do not require heating over 40°C, adding large volumes of diluent, or adding large volumes of chemicals to promote transportation through pipeline.

Methods

'EZ Flow' combines hydrodynamic cavitation processing of the heavy crude oil with addition of proprietary, low concentration chemical formulations (< 2% by volume) to promote the reduction of the heavy crude oil's viscosity. First, over thirty different chemical formulations were rapidly screened at the bench-scale with sonication and mixing as the mechanical mixing method. Second, the most promising chemical formulations identified at the bench-scale were scaled up to a small, 15-gallon pilot plant scale hydrodynamic cavitation system to process heavy crude oil and chemical formulation and to collect samples during and after processing for analyses.

Results

Chemical formulations 30, 31, and 32, treated using the 15-gallon hydrodynamic cavitation pilot scale system, yielded viscosity reduction of Belridge heavy crude oil (API° 11) by up to 60%.

Conclusion

'EZ Flow' reduced a Belridge heavy crude oil's viscosity up to 60%, making this technology a candidate for converting this type of heavy crude oil into a flowable material suitable for transport through existing pipelines at moderate temperatures. If commercialization continues as planned, the technology could result in significant decreases in transportation costs and potentially limit environmental impact for oil producers. There is also a high probability that this technology could be used for transporting other unconventional oils. Also, the treated heavy crude oil could be stored for extended periods of time (months) without losing its reduction in viscosity. EZ Flow technology is filed with the U.S. Patent Office in 2021, U.S. Patent Application Serial No.: 17/452,458.

639371 Design Criteria for Mitigation of Fouling in Reboilers

Abstract:

Significant progress has been made in understanding the underlying mechanisms of hydrocarbon fouling and characterizing the effects of physical and chemical parameters. However, the interactive effects of two-phase flows and fouling mechanisms are poorly understood. The performance of these exchangers directly impacts the overall performance of the process unit, such as distillations columns, hydrotreaters and also reactors. This is because design guidelines are inadequate to minimize localized process conditions that induce severe fouling condition, which could also lead to localized corrosion by concentrating corrosion species. This paper presents design criteria that should be followed to evaluate two-phase flow conditions under different operating conditions and design the exchanger to minimize severity conditions that would promote localized fouling under two-phase conditions. Such a design approach may not be optimum for heat transfer and pressure drop considerations under clean conditions, but it will minimize fouling and localized corrosion. The paper focuses on improved design of for minimizing fouling and under-deposit corrosion and a case study of revamping to minimize fouling propensity. The result is lower lifecycle ownership cost of the exchanger. Industrial heat exchangers in refining and petrochemical plants operate under multi-phase flow conditions, and fouling and localized corrosion are common problems in such heat exchangers. The major heat transfer equipment, not including fired heaters, in petroleum processing that encounter fouling and coking in a two-phase flow environment are: a) reboilers operating as thermosiphon or under forced feed conditions: localized fouling is generally caused by two-phase flow maldistribution on shell side and slug flow on tube side creating regional environment for chemical reaction fouling and particulate deposition of corrosion products that might be generated in the overhead condenser/separation drums, transfer pipelines, and/or in the distillation column; and b) feed/effluent (F/E) exchangers: F/E in hydrotreating and reforming processes are prone to fouling, mostly caused by chemically active species, such as iron-sulfide with non-uniform two-phase flows on shell side. The interactive effects of fouling precursors derived from blending products streams such as diolifins, reactive sulfur, dissolved metals reactive species like iron-naphthenate combined with two-phase flow pattern significantly enhances the fouling propensity. The chemical treatment to interrupt the interactive effects fouling precursors must take into account of two-phase flow for targeting these precursors. Chemical treatments developed using single-phase laboratory testing may not be effective in two-phase conditions. This presentation will provide a comprehensive approach of design criteria to follow for effective fouling mitigation by maintain the process parameters below threshold fouling conditions and effective chemical treatment.

640117 Testing to Improve the Reliability of Fouling Mitigation

Abstract:

Changing the operation and design of field heat exchangers involves the risk that a fouling mitigating action may not provide the intended benefit or potentially make fouling worse. For this reason, operators are often biased to replace bundles with the same design rather than use a new technology or design idea does not work out as planned. To help minimize risk, fouling testing can performed to evaluate the impact of proposed mitigation strategies and provide a data to make informed decisions. HTRI's high temperature fouling units are designed to perform fouling tests at the temperatures, velocities, and pressures of the hottest portion of the crude preheat train. There are seven test variables that may be leveraged to study fouling: bulk fluid temperature, initial wall temperature, flow rate, pressure, the fluid, the surface, and the test duration. Examples will be provided to illustrate how testing may be used to optimize design decisions as well as evaluate the benefit of proposed mitigation techniques.

639127 Anti-Fouling Ceramic Coating Technology for Heat Exchanger Tubes.

Abstract:

In many refinery units, such as Crude or Vacuum Distillation Units, high maintenance costs, furnace fuel consumption, heat exchanger pressure drop and unit throughput reduction are common problems. Fouling of the pre-heat train heat exchangers is one of the root causes that leads to these problems. Ceramic coating technology, minimizes fouling in heat exchanger tubes leading to significant improvements in heat transfer efficiency. The use of anti-fouling ceramic coatings in preheat trains leads to longer runs without cleaning, reliable operations, reduced fuel consumption and therefore significant OPEX savings.

639421 Optimizing Water Jetting Cleaning of Heat Exchangers

Abstract:

Cleaning of industrial equipment is a crucial activity that provides significant operational and economic savings. More specifically, the periodic cleaning aimed at the removal of fouling deposits in heat exchangers enables energy savings, debottlenecks production and contributes to reducing the carbon emission of industrial facilities. Cleaning of heat exchangers can be performed using a variety of methods (e.g. Mechanical, Chemical, Thermal, etc.) that can be selected depending on a number of factor such as equipment type, process fluids and conditions, fouling characteristic etc. Among the various options, Mechanical cleaning accounts for about 80% of cleaning actions performed in refinery and petrochemical plants, with hydro-blasting being the most popular amongst the mechanical cleaning methods. Hydro-blasting relies on high-pressure water jets that mechanically detach fouling deposits from the equipment, thus is often referred to as water jetting. The performance of this cleaning method mainly depends on the outlet pressure (pressure loss), stand-off distance, nozzle diameter, fouling deposit characteristics and pumping power. These variables directly affect the energy consumption (and related cost) of the cleaning process and its optimal value is of great importance for all cleaning technology providers. This work proposes an optimisation-based approach to find optimal operating conditions (namely flow rate, nozzle diameter, stand-off distance, fouling deposit properties and pumping power) that minimises pressure losses in the cleaning equipment, together with water consumption, fuel usage and related CO_2 emissions). This study provides a potential method for predicting the effectiveness of water jetting actions and achieving optimal cleaning performance with minimum energy consumption.

DE-RISKING THE USE OF COMPACT HEAT EXCHANGER DESIGNS THROUGH BETTER CLEANING

Roberto Tomotaki, Clean As New, Baytown, TX, USA
Byron Keiser, Tech Sonic, Morinville, AB, Canada

Introduction

Engineers are reluctant to place compact designs in even moderate fouling services because of the risk associated with "cleanability".

It is very common practice to find only straight-tube bundles in fouling services. These bundles are often designed with so-called "cleaning lanes", which are arrangements of gaps in the bundle that allow a clear path for hydroblasting jets through the bundle, and wider tube spacing to help prevent the entrainment of fouling within the bundle.

Designing for cleaning in this way reduces an exchanger's heat transfer surface area and thus its total heat transfer capacity, and increases the needed equipment footprint, network size and complexity.

The advent of cleaning methods, such as ultrasonic cleaning and thermal cleaning can relieve the design concerns by ensuring that even the most compact designs can be cleaned thoroughly on all its flow paths. The elimination of cleaning constraints will afford engineers the opportunity to simply and inexpensively improve heat transfer performance by replacing standard square-pitch bundles with higher performance compact designs. The added heat transfer performance may be used to improve process performance or increase maintenance intervals.

Compact Heat Exchanger Designs

Compact heat exchanger technologies provide an alternative to the typical shell and tube heat exchanger design. When contrasted to traditional straight shell and tube designs, compact exchangers are typically described as having a higher surface area to volume ratio, higher flow turbulence leading to higher heat transfer coefficients, and a higher flow shear stress leading to slower fouling build-up. Many of these types of exchangers can also be designed to having a true-counter flow between the cold and hot fluids producing additional heat transfer efficiencies. These advantages can potentially lead to a lower capital project requirement and higher energy recovery. Common examples of compact heat exchangers are welded plate exchangers, spiral exchangers, and twisted tube bundle exchangers.

A common apprehension of engineers in selecting compact exchanger designs is the concern about the ability to clean such units in cases where they become fouled. Even though fouling build-up is typically slower in compact exchangers, most eventually foul to the extent where a cleaning is required. These exchanger designs have narrow flow lanes where a hydrojet using a traditional hydroblast machine does to have a direct path to impact the foulant deposit. Hence significant foulant deposits may be inaccessible to traditional hydroblast techniques and are left behind after cleaning. These cleaning concerns are often based on a company's past negative experience with these exchanger technologies. The difficulty in cleaning has hampered the propagation of compact heat exchanger technology in the industry.

Even for traditional shell and tube exchanger, the question surrounding cleanability impacts design choices, often preventing the use of more compact designs. For example, U Tube bundle designs offer several advantages including: (a) The elimination of a floating head makes the installation simpler and less expensive, (b) Tube side cleaning is possible without bundle removal, (c) The design uses one half of the tube-to-tubesheet joints, reducing the risk of leak failures, (d) the design is generally less prone to thermal stresses, and (e) It is generally possible to have higher surface area within the same volume by eliminating the need for the floating head at one end. Despite the advantages, U-tube designs do suffer from a significant liability related to cleaning. The tube bends, particularly the tighter bends on the innermost tubes, can present a challenge to traditional hydroblast techniques, which can often fail, even damaging tube integrity, when trying to remove fouling from the bends. For this reason, U-tubes are often avoided in favor of traditional straight-tube designs.

Compact tube layouts are another way that higher heat transfer surface areas can be imparted in traditional tube and shell designs. The default design for fouling services is a wide, square pitch layout. This design sacrifices tube count for cleaning effectiveness with hydroblasting, providing open channels through the bundle through which hydrojets may be directed. Even with these "cleaning lanes" however, hydroblasting is unable to reach 100% of the tube outer surface with sufficient momentum to provide complete cleaning. Typical results from hydroblasting alone result in exchangers returned to service at less than 80% of design performance. This reality impedes any consideration of more dense tube layouts, such as triangular pitch, which offers no cleaning lanes for hydroblasting. Similarly, the adoption of twisted tubes and other modified tube designs to improve heat transfer has been hindered by concerns about foulant removal during cleaning.

Twisted tube designs (e.g. Twisted Tube® from Koch Heat Transfer) provide improved heat transfer through increased surface area and improved flow characteristics. Performance improvements over standard straight tube exchangers can exceed 40% in the same volume. These bundles however can present a challenge for traditional hydrojet cleaning when fouled.

Novel heat exchangers designs, such as welded plate exchangers provide very high performance in a small volume. These designs however are often not selected in fouling services because of concerns around cleaning effectiveness.

Better Cleaning Technologies

Fortunately, there are two well established technologies which have the potential to alleviate these concerns, providing the possibility to address the cleaning difficulties experienced with these compact designs. Ultrasonic and thermal cleaning technology are now industrially available and have shown promising results with many of these more efficient heat exchangers.

Clean As New provides ultrasonic cleaning and has successfully cleaned many types of compact heat exchangers in severe fouling services, including U-tube, welded plate, spiral, and twisted tube exchangers. Clean As New uses Tech Sonic's Ultrasonic technology, which works by allowing ultrasonic energy, in a large bath filled with suitable cleaning chemistry, to penetrate throughout the entire volume of a heat exchanger, loosening the fouling on both the OD and ID, which may then be effective rinsed away by hydroblasting. Normal hydroblasting techniques may be used on the ID to ensure perfect cleaning on the ID of any tube type or configuration. Because of the action of the ultrasonic bath to loosen the fouling, normal bundle blasting on the shell side is able to completely remove all fouling, including entrained fouling.

Figure 1. A heat exchanger being lowered into an ultrasonic bath (courtesy of Clean As New)

Heavily fouled welded plate heat exchangers are routinely cleaned to like-new performance in the company's Baytown Texas facility. The example pictures in Figure 2 show a welded plate exchanger from a "crude/resid" service which is restored to near 100% of clean design performance using a combination of ultrasonic cleaning and high-pressure rinsing.

Figure 2. Severely fouled welded plate exchanger, cleaned using ultrasonics. (courtesy of Clean As New)

Another example, Figure 3, shows an example of a badly fouled twisted tube bundle used in a naphtha cracker cleaned using the ultrasonic method. In total, 12 units were cleaned in four days, each unit being returned to like-new performance levels.

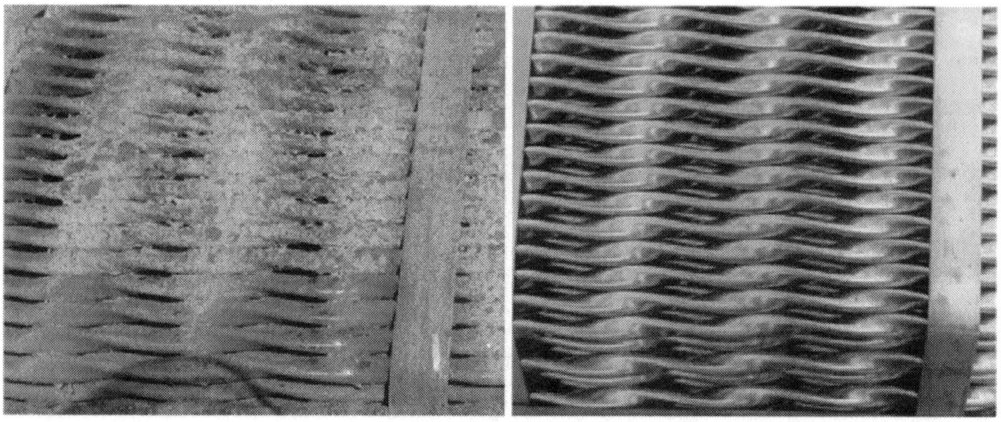

Figure 3. Twisted Tube® bundles cleaned using ultrasonic technology (courtesy Tech Sonic)

The following example, Figure 4, is a cleaning on a Crude Unit Atmospheric distillation overhead condenser. The condenser design tube layout was triangular. With this tube layout, cleaning lanes are not present. Traditional hydroblast methods is not capable of accessing the inner core of the bundle. Consequently, the client's distillation tower ran poorly. Using Ultrasound cleaning technology, the client was able to remove the foulant from the inner core of the bundle. The client reported being able to reduce the atmospheric tower pressure by nearly 50%, hence significantly improving the profitability of the Unit.

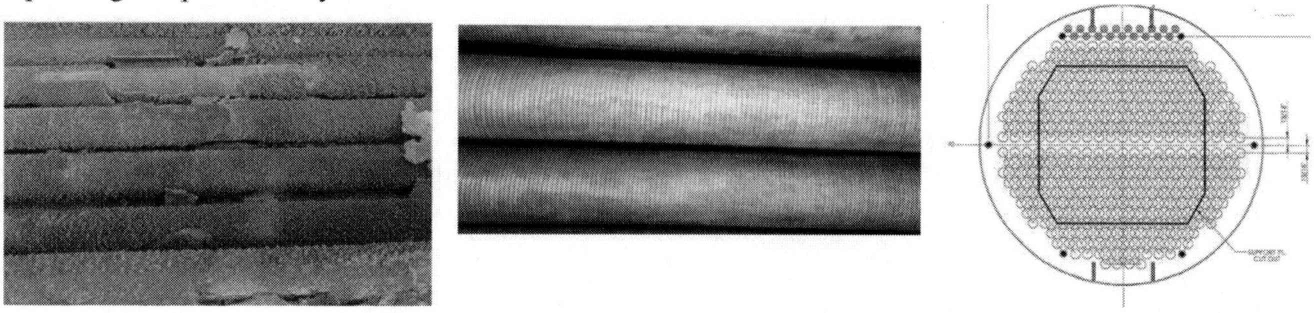

Figure 4. Triangular Layout Finned Tube Overhead Condenser cleaned using ultrasonic technology (courtesy of Clean As New)

Thermal cleaning offers an approach to cleaning compact heat exchangers fouled with materials that are not suitable for ultrasonic cleaning. Thermo-Clean, a Belgium-based company, has developed an advanced thermal cleaning system that is able to address difficult-to-clean hydrocarbon fouling, such as polymers, completely on even the most compact heat exchanger designs. Figure 5 shows a spiral heat exchanger, fouled with polyethylene. This exchanger can be completely cleaned of the polymer fouling by this technique, which under tightly controlled thermal and atmospheric conditions, will reduce any hydrocarbon fouling to ash, which is then rinsed away.

Figure 5. A spiral tube exchanger cleaned using controlled pyrolysis (courtesy Thermo-Clean)

A combination of ultrasonic and thermal techniques may be used when the fouling includes both polymer and inorganic material, such as scale on either the tube- or shell side. Thermal cleaning is not a suitable technique for inorganic fouling whereas most inorganic fouling may be removed quickly and completely using the ultrasonic method. In these cases, removal first of the inorganic scale by ultrasound followed by polymer removal by pyrolysis can restore even the most compact heat exchangers to 100% of design performance.

By demonstrating to the industry that these new approaches have the ability to clean compact exchanger designs in even challenging fouling services, and with a growing body of data to prove it, companies like Clean As New and Thermo Clean are easing the industry's concerns with using compact exchanger technologies in fouling services and offering new opportunities to improve performance

About The Authors

Roberto Tomotaki is a veteran of the oil and gas industry with over 25 years' experience in heat exchanger design, monitoring, innovation, troubleshooting, and repair. He currently serves as the Heat Exchanger Advisor for Clean As New and provides consulting services through Becht Engineering.

Byron Kieser has a 30 plus year career in academic research, teaching, chemical instrumentation business development, marketing, and ultrasonic technology development. He is the Director of R&D for Tech Sonic.

639530 Matching the Cleaning Method to Your Operational Goals.

Abstract:

There are many choices available when it comes to the cleaning of heat exchangers. Over the last number of years, there has been much discussion about matching your cleaning method to the type(s) of fouling present within the heat exchanger in question. Great approach. But what if your present set of circumstances does not allow you to choose the method best suited to removing the fouling?

This presentation looks at this very question and takes it a step further. We will review how goals surrounding: operation, asset longevity and cost avoidance, can often play a major factor in the decisions we make, for cleaning heat exchangers.

This presentation will talk to the latest advancements in ALL cleaning categories. We will walk the audience through a few industry case examples, where combining particular cleaning methods have led to the achievement of both operational goals and key cleaning objectives.

About the author of this presentation:

Using the ORANGE technology to optimize heat exchanger performance, Russell Philion, often works with a broad spectrum of industrial cleaning vendors, each offering differing methods for cleaning heat exchangers. This working knowledge of the various cleaning methods combined with data gathered from end-clients and key refinery personnel, have provided the foundation for this subject material.

640008 "Liquid Entrainment in Kettle Reboiler with Asymmetrically Located Nozzles"

Abstract:

Entrainment is the carryover of the liquid droplets in vapor streams from kettle reboiler and can cause low efficiency in the distillation column. In the other applications for kettle type exchanger as a vaporizer , even a small amount of entrainment entering the rotating equipment such as centrifugal compressors and pumps will increase power requirements and damage rotating elements. These type of issues during plant operation can be avoided with the proper thermal design of kettle reboiler to reduce entrainment ratio for the stream leaving exchanger.

Sufficient disengagement space should be provided in a kettle reboiler to minimize entrainment of liquid and increase efficiency of the equipment. The amount of entrainment depends on:

- Kettle size
- Number of nozzles
- Vapor generation in the bundle
- Vapor space above the bundle
- Use of demisters

HTRI is the common software for thermal design of kettle reboilers to predict entrainment ratio considering different parameters.

Presentation will focus on Liquid Entrainment in Kettle Reboiler with asymmetrically located nozzles to show how entrainment ratio will increase when nozzles are located asymmetrically.

HELIXCHANGER® Heat Exchanger

*S. Taha[1], R. J. Jibb[1], R. Groppi[2]
[1] Lummus Technology LLC, Bloomfield NJ
[2] Lummus Technology Heat Transfer BV, The Hague, Netherlands

Abstract

Lummus Technology provides a look back over the past 20 years in commercializing the HELIXCHANGER® heat exchanger technology. The paper dives into how the technology has improved over the years in solving vibration risks to increasing run-lengths and reducing total operating cost (OPEX) for certain fouling or viscous fluids. Guidelines have been developed per application type to provide insight into whether and where HELIXCHANGER helical baffles can provide an optimal solution. Finally, we look towards what improvements are being developed and where the HELIXCHANGER® heat exchanger technology is heading.

Introduction

With over 5000 units in operation worldwide, the most widely used helical baffle geometry in industrial applications is the HELIXCHANGER[1] heat exchanger. The HELIXCHANGER technology was developed from the early design of Lutcha and Nemcansky [1] in The Czech Republic and later acquired by Lummus Heat Transfer at the end of the 90's. The design was improved further as described in US Patent 6827138 [2].

Figure 1 HELIXCHANGER heat exchanger

In this paper we summarize the basic geometrical details of the HELIXCHANGER baffle design, and provide examples of some of the most important core applications that have emerged since the

[1] *HELIXCHANGER® heat exchanger is a proprietary shell and tube heat exchanger technology licensed by Lummus Technology LLC*

technology was first introduced. A recent case study is provided comparing performance of a HELITOWER™[2] design to a conventional No Tubes In Window (NTIW) style Texas Tower. Finally, we describe some recent work to validate a Computational Fluid Dynamics (CFD) model of a HELIXCHANGER bundle with laboratory data obtained from Lummus ongoing experimental program.

HELIXCHANGER Basics

The term "Helical Baffles" refers to a class of baffle structures used in shell and tube heat exchangers to direct flow on the outside of a tube bundle in a primarily helical path around a central core. This contrasts with flow in shell and tube heat exchangers with cross-baffles (predominantly crossflow, perpendicular to the tubes) and "grid" or "rod" baffles (predominantly longitudinal flow, parallel to the tube axis). The HELIXCHANGER baffle has the following features:

a) Quadrant shaped baffles positioned at an angle to the tube axis along a central rod.
b) Baffles act as guide vanes rather than flow barriers.
c) Baffles are cut from an *elliptical plate* so that the clearance between the baffle and the shell is maintained constant following, for example, TEMA standards.
d) Each baffle quadrants *overlaps* with the adjacent baffle quadrant creating a balance between the rotating helical component of the flow in the outer cylindrical region and the higher velocity vortical flow in the center of the bundle. The optimal combination of the *baffle angle*, and *overlap* will yield a uniform temperature profile and minimize axial dispersion.
e) The overlap also allows a reduced un-supported tube span.

When compared to a conventional shell & tube exchanger with segmental baffles a HELIXCHANGER heat exchanger can provide:

- **Higher average shell side velocity** for the same allowable pressure drop. The higher shell side velocity can be beneficial in improving the shell side heat transfer coefficient.
- **Higher shear** rate at tube walls reducing shell side deposit fouling.
- Shell side flow volume with virtually **no "dead zone"** where fluid can stagnate or back-mix.
- A **shorter unsupported tube span** compared to segmental baffle heat exchanger reduces the risk of flow induced vibrations

HELIXCHANGER bundle Geometry

Referring to Figure 2, baffles are connected to a central pipe or rod of diameter D_r at an acute angle ϕ_B to a plane perpendicular to the central axis of the shell. Each quadrant may be labelled from A through D clockwise around the shell with A at the 12 O'clock position. The baffles are arranged along the central pipe with a spacing B between adjacent baffles in the same quadrant.

A limiting case is defined for adjacent baffles touching at the perimeter such that a helix is described on the inner surface of the shell by the curved edge of the baffles. The maximum possible baffle spacing B is approximated by the helical lead H_S, where:

---

[2] *HELITOWER heat exchanger refers to vertical feed effluent heat exchanger with HELIXCHANGER baffles*

$$H_S = \pi D_S \tan \phi_B \tag{1}$$

Typically, the baffles in the same quadrant (A through D) are spaced apart by a distance B (where $B < H_S$) with a corresponding crossing fraction γ.

For $B < H_S$

$$B = \gamma \pi D_S \tan \phi_B \tag{2}$$

In fact, since the baffle plates themselves are flat, the path described within the tube bundle differs slightly from the elliptical path defined by Equation 1, An alternative dimension for the lead can be defined which is not a true helix, but accounts for the flat surface of the baffle plates

$$H_{S2} = 2\sqrt{2} D_S \tan \phi_B \tag{3}$$

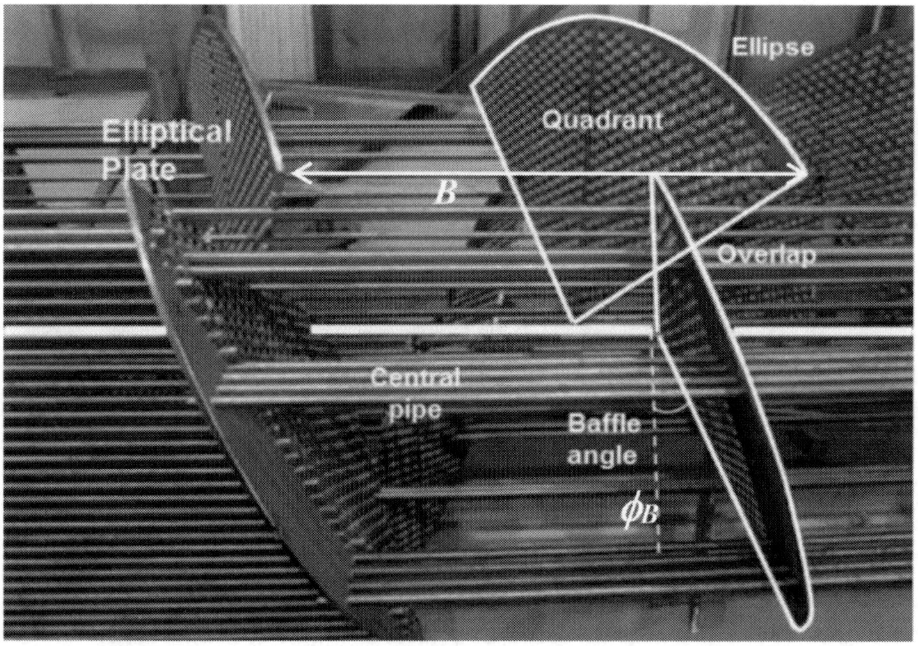

Figure 2 HELIXCHANGER baffle layout (tubes not installed)

The fluid path through the bundle is complex, combining both rotational flow and longitudinal flow. The helical baffles separate the flow above and below the crossover point and the fluid describes a pseudo helical path through the tube bundle whereby the average axial distance of the fluid flow per rotation is equal to or greater than B but less than H_S.

HELIXCHANGER heat exchanger applications

For a given shell diameter, helical flow can provide a higher average velocity than crossflow type segmental baffles whilst maintaining similar pressure drop. This can be an important consideration in many fouling applications where high shear stress on the external tube surfaces is desirable to limit fouling, or for viscous fluids where higher shear rates can significantly improve performance. Barletta and Chunungad [3] have pointed out that HELIXCHANGER technology can be applied to crude pre-heat trains to increase both shell-side and tube side velocity by enabling a smaller shell diameter than can be achieved with segmental baffles within the given constraints. In other words, the use of HELIXCHANGER technology can provide a design envelope that allows for higher velocities on both shell-side and tube-side. They suggest that fouling rates can be reduced by maintaining shell-side velocities of 1.2 – 2.4 m/s and tube-side velocities of 2.4 m/s or higher which will offset initially high pumping costs by reducing pressure drop associated with fouled units.

The helical baffle construction also provides a shorter unsupported span than segmental baffles which in turn reduces the risk of flow induced vibration at higher velocities. Where vibration concerns or available shell-side pressure drop are limiting, it is possible to vary the baffle angle, spacing and external sealing in order to maximize shell-side velocities.

Table 1 summarizes some of the core applications where HELIXCHANGER technology is widely used. References to pre-existing published case studies are also provided where applicable.

***Table 1** – Core applications of the HELIXCHANGER heat exchanger Technology*

Goal	Example applications		HELIXCHANGER Heat exchanger benefit
	Shell Side Fluid	**Service**	
Reduce shell side fouling rate	Crude	Crude preheat trains **[i]**	Helical flow pattern has been proven **to reduce fouling rates** through combination of lower tube surface temperature and an increase in the shear stress over the tube surface. The continuous helical flow pattern reduces dead zones and back mixing.
	Vacuum Residue	Vacuum Residue exchangers **[ii]**	
	Quench water	Quench water coolers (Ethylene)	
	Charge Gas	Charge gas aftercoolers (Ethylene)	
	Cooling Water	Butadiene Column Condenser	HELIXCHANGER bundle enables a **higher shell-side velocity** within the allowable pressure drop which is desirable when it is necessary to place cooling water on shell-side. Also a HELIXCHANGER bundle provides better drainage when exchanger is installed in vertical position. The higher shell side velocity may allow reducing cooling water flow rate requirement
		Propyne Column Condenser	
		Afterwasher Condenser	
		Charge gas aftercoolers (PDH)	
		Effluent coolers	
Prevent flow induced vibration	Propylene Refrigerant	Propylene Refrigerant condenser	The **reduced unsupported span** reduces the risk of flow induced vibration. This allows to convert NTIW design to HELITOWER design and fit more tubes in the same shell ID or reduce shell ID for same heat transfer area.
	Propylene	Propylene fractionator condenser	
	C4 Hydrocarbons	Main washer condenser (Butadiene)	
	Hydrocarbon/ Hydrogen	Feed/Effluent tower (HELITOWER) **[iii]**	
Effectively heat and cool high viscosity fluid on shell side	Sales Oil	Sales oil cooler **[iv]**	HELIXCHANGER overlapping baffle design rotates and divides the flow, in a similar way to a static mixer device to **more effectively heat or cool viscous fluids**, or fluids which have a large range of viscosity with temperature.
	Vacuum Bottoms	Vacuum Bottoms/Crude Exchanger	
	Vacuum Residue	Vacuum Residue Cooler **[iii]**	
	Emulsion	Emulsion Exchangers	

Published reports which provide case studies and operational feedback for the applications listed above.

[i] Master et al. [4,5] have presented several case studies covering refinery applications in crude preheat and feed-effluent services. In one example, an increase in run length from 1 year between cleanings to over 2 years in *crude preheat* service is reported. Gough et al. [6] have provided a case study for grass roots design case study combining HiTRAN® and HELIXCHANGER technologies for a crude oil interchanger on a floating production storage and offloading ship platform refining process. The advantages were smaller plot space at lower energy cost with reduced fouling tendency as compared to conventional exchanger design. The helical flow velocity was more than three times higher than the average cross flow velocity achieved in the equivalent segmental-baffle design.

[ii] Jibb et al. [7] have presented three case studies which illustrate improvements in run length for heating heavy fouling fluids such as *vacuum residue*. Significant increases in run length are again reported.

[iii] van der Ploeg and Master (1996) provide an early description and case study for use of helical baffles in a Continuous Catalytic Reformer (CCR) feed/effluent application. HELITOWER Feed Effluent heat exchangers have since been used for Lummus OCT and DPG, UOP Platformer, Axens Reformer, Exxon Powerformer and MTPX, Haldor Topsoe HTU and HDS

[iv] Jibb et al. [8] have described a 2018 study whereby HELIXCHANGER bundles were installed in one of two parallel trains used to cool a viscous sales oil. The sales oil bulk viscosity ranged from about 20 cP at the inlet to about 300 cP at the outlet. The performance of the HELIXCHANGER bundles was 30-50% higher than the parallel train using the original conventional baffle design (vertical cut single segmental baffles). The HELIXCHANGER baffle design was able to eliminate the geometry induced maldistribution, which allowed the heating section to be operated without addition of diluent.

Case study: HELITOWER.

A North American operator requested Lummus Heat Transfer to provide a replacement for an existing vertical feed effluent heat exchanger (Texas Tower) with a condensing effluent stream (Naphtha and Recycle gas) on the shell-side and a boiling feed (Naphtha and Recycle gas) on the tube-side. The replacement HELITOWER was designed to have the same overall dimensions as the existing unit ("Old Texas Tower") but with larger tubes for easier tube-side cleaning. Table 2 provides the geometry of the existing unit ("Old Texas Texas tower") and the replacement HELITOWER.

Table 2 *Dimensions of existing unit ("Old Texas tower") and replacement HELITOWER.*

		Old Texas tower	New HELITOWER
Size	mm	1760 x 18300	1760 x 18300
Tube OD	mm	15.875	19.05
Tube pitch	mm	22.25	25.4
Tube layout angle	deg	45	45
Tube Count	-	3510	3100
Area	m2	3150	3316
Baffle type	-	NTIW	Double Helix
Special requirements	-	Deresonating baffle	None

The tube diameter for the HELITOWER design was increased from 15.875 mm to 19.05 mm, which provided improvements for tube-side cleaning access. For the HELITOWER design, the full shell volume could be used, and a deresonating baffle was not required. The resulting surface area was ~5% higher than the previous design, based on a better use of available volume despite using larger diameter tubes.

Performance was monitored before and after the HELITOWER replacement, for almost identical feed inlet conditions over a period of 7 months (for the HELITOWER replacement unit). Table 3 provides the performance of each geometry with the very similar inlet operating conditions, but with a hot end approach of 49 C for the HELITOWER replacement compared to 64.5 C for the original Texas Tower.

Summary

- HELITOWER technology more effectively utilizes the available shell volume than the prior NTIW arrangement, without risk of flow induced tube vibration.
- The tube diameter was increased from 15.875 mm (NTIW Texas tower) to 19.05 mm (HELITOWER) with still 5% more surface area in the same shell.

- The hot end approach was reduced from 64.5 C to 49 C, with a 5% increase in total heat duty at the same feed flow rate.

***Table 3** – Performance data before and after HELITOWER replacement*

		Old Texas tower	New HELITOWER
Data point		21-JUN-17 15:18:00	6/28/18 12:00 AM
Flow rate	BPD	12,268	12,501
	kg/hr	61,185	62,348
Feed in	C	56.30	56.40
Effluent in	C	488.70	486.80
Feed out	C	424.20	437.80
Effluent out	C	113.90	102.80
MTD	C	52.60	40.90
Hot end approach	C	64.50	49.00
Average duty	MW	23.00	24.25

Computational studies of HELIXCHANGER baffles

Andrews and Master [9] used the CFD code HEATX to carry out three-dimensional modelling of a HELIXCHANGER heat exchanger. The model used a porous medium to describe the tube bank. The helical baffles were represented as porous surfaces to account for the leakage between the tubes and the baffles. This technique enables a CFD simulation to be done with significantly fewer cells than if the tubes and gaps were resolved with a mesh, but must be carefully calibrated using experimental data or sub-models for the flow across the individual tubes. The study concluded that the porous media approach provided reasonable values of the shell side pressure drop. However, no comparison with experimental data was provided so it is not possible to determine the accuracy of this model.

Jibb et al. [10] have described a methodology for pressure drop prediction that is more applicable to larger scale industrial HELIXCHANGER units, with gas flow at higher velocities. The method uses a STAR CCM+ CFD method using a periodic approach, accounting for the tube to baffle clearance. The model agreed with available field data to within 5.8%. A model of a HTRI test Unit 23 (conventional baffled heat exchanger) under predicted measured pressure drop by 3 -21%. Finally, a test rig was built specifically for the testing of pressure drop across HELIXCHANGER heat exchanger bundles, and data was generated over a range of flow rates. The CFD model agreed with this data to within -8% to +0%.

In the present work, data obtained from a HELIXCHANGER test loop was used to validate a CFD modelling approach to simulate heat transfer and pressure drop for HELIXCHANGER heat exchangers with water on the shell side. The test article was a TEMA BES type bundle with a 7 degree baffle angle.

Figure 5 – HELIXCHANGER Test Facility and test bundle

The test article was modelled using FLUENT software. In order to reduce computational time, the exchanger a porous medium domain was defined to represent the tube bundle, as shown in Figure 5. Whilst a more explicit model was technically possible for the smaller test article, the porous medium approach was selected because the calibrated and validated CFD method could subsequently be extended to full scale industrial bundles, which would be too large to model explicitly considering all of the tube to baffle clearances.

Explicit sub-models were created to determine the heat transfer and pressure drop of the porous domain in terms of (i) resistance coefficients for the tube field, (ii) resistance of the baffles with associated clearances and (iii) e-NTU relationship for heat transfer between the tubeside and shellside fluids.

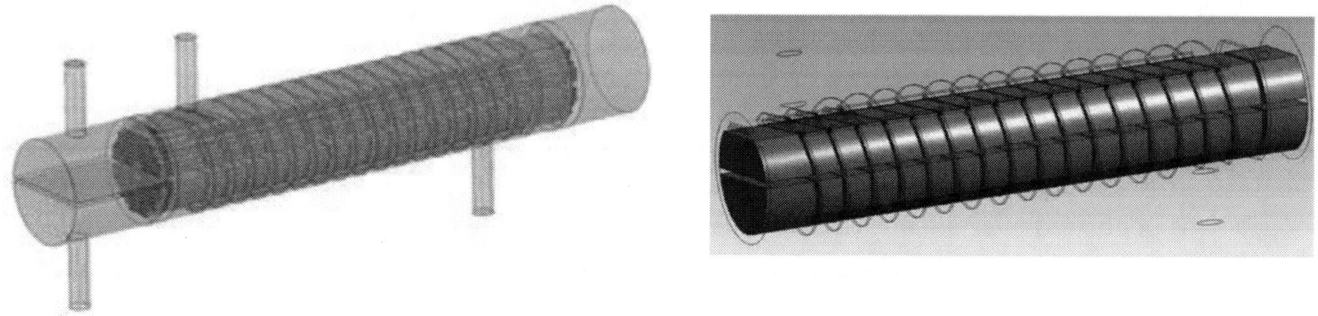

Figure 6 – *Porous Medium model of HELIXCHANGER test article*

The model was used to predict outlet temperatures and pressures based on defined inlet conditions for three experimental runs with a range of shellside Reynolds numbers (based on average helical flow velocity) between 18,000 and 40,000.

Table 4 – *Comparison between experimental data and CFD Porous Media predictions*

Case Identifier	Case 7	Case 13	Case 14
Shell side Reynolds number	40,000	26,000	18,000
Nozzle to Nozzle Pressure Drop (Expt – CFD) %	-3.7%	-5.2%	-6.4%
Bundle pressure gradient (Expt – CFD) %	2.3%	-0.7%	-2.4%
Heat transfer rate (Expt – CFD) %	2.6%	-0.8%	-12.8%

The predictions from the CFD Porous Media model were found to be in good agreement with the experimental measurements for turbulent shell-side flow for both heat transfer and pressure drop. This result provides a basis to use the model for larger HELIXCHANGER bundles used in industrial services.

Conclusions

- HELIXCHANGER heat exchanger technology is proven technology which can be used in grass root project or revamp projects to improve heat exchanger performance.
- HELIXCHANGER heat exchanger design option should be investigated whenever shell side fouling or flow induced vibration is an issue. Some examples of these core applications are given in Table 1.
  - Higher shell side velocity achievable is a HELIXCHANGER heat exchanger compared to segmentally baffled heat exchanger can significantly increase heat exchanger run-length reducing the operational cost (OPEX).
  - Improved shell side heat transfer coefficient and/or superior flow induced vibration resistance can reduce the heat exchanger size reducing the capital cost (CAPEX).
- The HELITOWER technology has been applied to replace large vertical feed effluent heat exchangers. In the case study presented, a 5% increase in surface area was possible despite a requirement to increase tube diameter. The performance of the HELITOWER was successfully demonstrated.
- Computational fluid dynamics has been successfully validated against experimental heat transfer and pressure drop data for turbulent shell-side flow. The validated modelling approach will be used to predict performance of larger HELIXCHANGER bundles used in industrial service.

References

[1] Lutcha, J., and J1 Nemcansky. "Performance improvement of tubular heat exchangers by helical baffles." Chemical Engineering Research and Design 68, no. 3 (1990): 263-270.

[2] Master, Bashir I., Krishnan S. Chunangad, and Venkateswaran Pushpanathan. "Heat exchanger." U.S. Patent 6,827,138, issued December 7, 2004.

[3] Barletta, T., White S., and Chunangad K.S., "Designing a crude unit heat exchanger network." www.eptq.com Sour and Heavy Petroleum Technology Quarterly 16 (2012): 3.

[4] Master, Bashir I., Krishnan S. Chunangad, and Venkateswaran Pushpanathan. "Fouling mitigation using HELIXCHANGER® heat exchangers." (2003): 317.

[5] van der Ploeg H.J., Master B.I., "He;ical Baffled Heat Exchangers – Design Option for Reactor Feed/Effluent Heat Exchange", Heat Exchange Engineering Conference, Birmingham UK, Advances in Industrial Heat Transfer IChemE (1996)

[6] Gough, M. & Dröegemüeller, P. & Ellerby, P. & Lee, M. & Master, B.I. & Chunangad, K.S.. (2007). Dual enhanced heat exchangers using Cal Gavin hiTRAN® wire matrix tube inserts along with the ABB-LGHT HELIXCHANGER® technology. 17. 32-41.

[7] Jibb, R. J., M. Brignone, S. Taha, G. Podrebarac, J. Larance, and P. Junge. "HELIXCHANGER® HEAT EXCHANGER-FIELD EXPERIENCE AND ONGOING DEVELOPMENT." Heat Exchanger Fouling and Cleaning XIII 2019, 2-7 June 2017, Józefów (Warsaw), Poland.

[8] Jibb R., Helixchanger® Heat Exchanger - Case Studies and Experimental Program, (68c), AIChe 2020 Virtual Spring Meeting and 16th GCPS 2020 Paper 68c.

[9] Andrews, Malcolm J., and Bashir I. Master. "Three-dimensional modeling of a Helixchanger® heat exchanger using CFD." Heat transfer engineering 26.6 (2005): 22-31.

[10] Jibb, R., Taha, S., Ebeling C., Gallei, W., Macedo E., "HELIXCHANGER® Heat Exchanger – Experimental and Field Pressure Drop Measurements used to validate CFD model", Heat Exchange Engineering 2014

Appendix: Comparison between experimental data and CFD Porous Media model predictions

			Test Rig	CFD	
Case 7 - Re = 40130	Shell side inlet Temp.		67.5	67.5	C
	Shell side outlet Temp.		52.3	51.8	C
	Tube side inlet Temp.		16.7	16.7	C
	Tube side outlet Temp.		35.6	36.2	C
	Duty		394	407	kW
	Shell side DP	Noz-Noz	4.23	4.08	psi
		Port 1-13	1.48	1.52	psi
Case 13 - Re = 26740	Shell side inlet Temp.		61.6	61.6	C
	Shell side outlet Temp.		43.2	43.3	C
	Tube side inlet Temp.		8.3	8.3	C
	Tube side outlet Temp.		25.7	25.6	C
	Duty		358	356	kW
	Shell side DP	Noz-Noz	2.47	2.34	psi
		Port 1-13	0.89	0.88	psi
Case 14 - Re = 17840	Shell side inlet Temp.		64.6	64.6	C
	Shell side outlet Temp.		40.2	43	C
	Tube side inlet Temp.		8.8	8.8	C
	Tube side outlet Temp.		24.2	22.4	C
	Duty		314	279	kW
	Shell side DP	Noz-Noz	1.15	1.08	psi
		Port 1-13	0.42	0.41	psi

639571 Effective Troubleshooting of Shell-and-Tube Heat Exchangers Using CFD

Abstract:

The complex physics inherent in modern multiphase process heat exchangers, driven by increasingly lower temperature approaches and high effectiveness targets, mandates that troubleshooting skills be comprehensive and expedient. While the modeling capability, fidelity, and execution speed of computational fluid dynamics (CFD) continue to evolve, solution times and the requisite pre- and post-processing activities still generally exceed the available time frame for the engineer urgently diagnosing a problem exchanger in the plant. However, various modeling approaches can significantly decrease turnaround times for CFD studies. Porous media modeling can accelerate CFD solution times and maintain high accuracy via the *f*- and *j*-factors vs. pitch and layout angle relationships HTRI has developed over many years of industrial R&D. CFD can even efficiently model exchangers with phase change, like reboilers or condensers, when coupled with HTRI's ***Xchanger Suite***®. Computing the phase change fundamentally in the CFD simulation substantially increases the complexity and duration of that simulation with dubious accuracy. The spatial distribution of vapor sources (reboiler) or sinks (condenser) in the porous medium (representing the tube bundle) can be input to the CFD simulation from the output of HTRI's control volume analysis in ***Xist***®, which is backed by over five decades of empirical data and analysis. The synergistic use of HTRI's control volume, empirical-based approach for the thermal behavior with a traditional CFD approach for the hydraulic behavior provides an accurate and timely solution that can impact troubleshooting activities. To demonstrate the utility of this approach, we present a few case studies concerning thermal performance and flow-induced vibration issues.

639041 Optimization Concepts utilizing Welded Plate Heat Exchanger Technology – Doing more & better with less!

Abstract:

What if you could reduce the size of your distillation condenser while achieving a closer temperature approach, resulting in not only greater energy savings but water and installed cost savings as well? In the last two decades, the design, optimization, and implementation of welded plate heat exchanger technology has leap frogged the conventional tubular heat exchanger technology design & optimization methodology. The results? More improved process design, compact equipments, lower capex and lower energy and water usage. You'll also get a more modular structure with design flexibility built in.

In this session, understand few concepts & ways of optimizing your process & heat exchanger design utilizing welded plate heat exchangers barring old rule of thumb which can positively impact not only the exchanger itself but overall unit operations including air coolers, trim condensers, compressors, reboilers, structure, column performance, overall footprint, CAPEX, and project ROI.

AUTHOR INDEX